United States and Mexico

A Border Proposal to Benefit Both Nations

Clifton Ray Wise Ed.D.

Author: Clifton Ray Wise Ed.D.

Epub : 979-8-9899236-4-9

Paperback : 979-8-9917742-9-1

Publisher: Clifton Ray Wise

ACKNOWLEDGEMENTS

We like to think we are *self-made*, but in reality, no human entity is self-made, we are products of our environment and both positive and negative relationships, from when we are first brought into the light at birth until the day we draw our last breath and move to darkness.

First I would like to thank my brain for never seeming to stop coming up with ideas to help my students and my fellow man. I would also like to thank artificial intelligence (AI) for coming up with the statistics and geographic portion of this writing.

P R E F A C E

It would seem to me that instead of a simple fence as a barrier that there could be a more productive barrier put into place that would benefit not only my country the United States of America but our physically attached neighbor to the south, Mexico. We hold an exceptional alliance with our attached neighbor to the north, Canada, and I do not see why we cannot have a similar position with our southern counterpart.

This writing attempts to propose an idea that would create a massive undertaking but nothing that cannot be attained when you have the brain power of the human entity.

I do not mention this idea for the purpose of entertaining another trade route to bring foreign goods into the United States as I think we have enough of that already, and need to rethink our current situation on that matter.

I bring this idea forth as I see the potential for both nations to create much needed:

- Much needed employment positions for low and middle income citizens.
- Bringing water resources to land that has and will never see those resources.
- New industry to convert gulf water to potable to feed agricultural land that have never seen that potential.
- Miles and miles of development potential for waterfront resorts and activities.

- Putting an abrupt stop to current drug trafficking through tunnels into the US to our border cities.

These are just a few of the reasons I have proposed this idea.

Throughout our history the United States has been the sole entity that has looked out for and assisted all other nations during their time of strife and need. Our unparalleled humanitarian country needs to take time to improve our country to make it safe for our population and the other creeds and populations that want to visit and transfer here.

Table of Contents

INTRODUCTION

The concept of my border idea is my sole idea through critical, creative and strategic thinking. I have attempted to provide a useful suggestion and propose it as an attempt at bringing two nations together through a participatory challenge: United States of America and Mexico.

1 ... P R O P O S E D   B O R D E R   I D E A

I'm not going to go into the political issues relative to our United States border and the security it either or neither proposes to provide.

Typically when you are involved in security in some way, shape or form you come to the realization first that there is no such thing as 100% safe and secure and second that the level of security you achieve, think you have achieved, or seek to achieve boils down to how much money you are willing to provide to that venture.

In general, that is the way it goes although there are some instances, as there are always exceptions to the rule. For example: If you ride a motorcycle? How much money do you propose to invest in a helmet? Do you propose the $50 helmet of the $400 helmet? If you glide on the tar or concrete how much does a pretty face mean to you? I assure, I have seen them ground down and it isn't a pretty sight. You could also become paralyzed from an accident regardless of whether the helmet was intact or not. Like I said, exceptions to every rule, but you have to look at the average or general idea or concept. That is where the real truth may lie. The price of the helmet is the price you are willing to pay for the perceived security you need to make you feel safe and secure.

Our US Border

December 11th of 2018 I wrote a letter to our current United States President at that time, Mr. Donald Trump.

I know, do I really think he would get a letter from me, just a simple taxpayer? I was hopeful, but I doubt this really happens like in the movies. It was probably just a waste of time, but I had something to say, an idea, and I got it out of my system regardless of who read it or not. Here is the idea I had and the exact wording of the letter I sent:

"An Idea about Our Border Security with Mexico"

Dear Mr. President:

I am a professor at Kentucky State University. This message is not relative to my position or standing nor even my teaching expertise area. Although, problem solving, critical and creative thinking are areas in which I have knowledge and expertise. We have heard mention many times of the necessity of building a barrier between our country and our neighbor to our south, Mexico. The barrier that is often mentioned is that of a wall between our two countries. Why not "think outside the box?" A barrier could come in many shapes, sizes or forms. This is why in the following paragraphs I mention a barrier that would consist of a canal or waterway between our two countries.

United States presidents have been involved in major economic construction projects in our past from: Park and Recreation, Inter-State Highway, and Valley

Authority Dam projects. In addition, the Panama Canal was an extremely large undertaking that took 10 years to complete and it was a stretch of only 48 miles. Of course, with the technology of today that project would take far less than 10 years for completion. These projects have brought resolve to many issues surrounding our country at a time when the need happened to be great and these projects were successful in many ways. Your stamp on an undertaking as this would mark your legacy within our American history.

The pros of a canal: Economic prosperity in the form of jobs for both countries; Dredging a waterway would collapse drug trafficking tunnels between our countries; Water would be brought to areas that have never had water before; Building of salt water to fresh water conversion industry in the area; and new Coastal/Waterfront real estate for both countries. Additionally, there will be a mass of sand, dirt and rocks that can be transported throughout our gulf states to reinforce beaches and barriers designed to prevent disasters.

I did not add cons as I see no actual cons when looking at this concept. Both of our countries would benefit from this type of barrier. If we need a barrier, why not make it more than just a barrier, make it a useful barrier for all parties involved. Both countries could be shareholders and be proud of what they have achieved.

I would only ask that a feasibility study might be in order, if this barrier idea might be of interest to you. In addition to you, I would like to see "America Great Again."

Thank you for your time.

Respectfully,

Clifton Ray Wise

This is what I received back:

Thank you for taking the time to express your views regarding the construction of a wall along our southern border.

The high volume of illegal immigrants flowing across our southern border continues to pose a significant threat to our citizens. Federal agents apprehend hundreds of thousands of individuals who are in our country illegally each year, many of whom are members of dangerous gangs that have wreaked havoc in their home countries. Transnational criminal organizations actively run complex drug and human-trafficking schemes, contributing to a significant increase in violent crime and death from dangerous drugs in the United States. This is unacceptable. We need a border wall to stop the trafficking of drugs, weapons, and people.

Shortly after taking office, I signed an Executive Order calling for the planning, design, and construction of a border wall. Earlier this year, I personally visited eight border wall prototypes, each of which stands approximately 30 feet high and has effective anti-breaching and anti-climbing capabilities. My Administration continues to press Congress to establish a $25 billion trust fund for the border wall system.

Thank you again for writing. To learn more about the actions my Administration is taking to secure our borders, please visit www.WhiteHouse.gov/Borders. As President, I am committed to keeping my promise to build a wall that secures our southern border.

Sincerely,

I have always been curious about our border with Mexico. I have always wondered if the things we see on television are real of just staged? In the world of today, where are the un-bias news reporters of yesteryear. Where are the Walter Cronkite's and the Tom Brokaw's? Where have they gone? It seems as though every news entity today is owned by someone who has a political affiliation and the reporters and news they project lean toward their bias.

The more I thought of our border with Mexico and the issues there I had to look back at the start of all this immigration. We would have to go back to the time after the Great Depression in the US. There were quite a few family farms that came back after the dust bowl and there were others that bought land out and started farms. Quite a bit of this was during some of the US Presidential ventures they proposed to bring the US economy back, like the Tennessee Valley Authority with building river dams, the US Interstate Highway System building state-to-state city-to-city roadways, and Federal State Park systems. All things made to give the public population jobs. Quite a few of these jobs were lucrative for the times and paid a somewhat decent wage and sometimes benefits, which were also new to those times made through political legislation.

However, these jobs were lucrative, there was a workforce need for the more menial labor les-paying related jobs and that was of working in fields of crops on farms. At that time there were smaller family farms then the corporate entities became involved and bought farmers out and made industrial sized corporate farms. Well, all of these farms had the same issue, finding workers to fill the slots to make production. What did they do for workers? Since, at that time there were no laws against borders or wage implementations or impositions that we see today. Farms hired the people that would do the work. The immigrants that just wanted to survive and make a living. Farms knew once they got them to their farms they were misplaced and could be taken advantage of and that is what happened and I read continues to happen even at this time. Farms brought them here because they could get them cheap.

So Here They Are

So now what do we do with them? In elementary public education we are told the United States is derived from all races from all regions of the world, a melting pot. I guess it all boils down to: If you want to come to the USA and get a job and pay taxes it is fine. I would guess? Someone has to do the jobs others don't want to do or in fact will not do! How do those jobs get done? They don't magically get resolved. Or do they? I've never seen a magic wand.

I had a friend come up from Texas last year and stayed a few weeks. He was of Mexican descent. The town he lived in was close to the US Mexico border. He would talk about how the US news was always rigged and staged for the public and that is was not really a true picture of the reality. He pointed out the political sides and how those staged locations were nothing more than a facade. The political figures would go to the somewhat cleaned-up spots and you never would see the reality of the dead bodies lying on top of each other or the semi-truck trailers filled with bodies the human traffickers would leave behind in the middle of nowhere to virtually bake in a trailer over with no water for days, until someone would finally open the doors and find them all dead. He said it was a real travesty that the people of the USA was not really given all the facts.

My Border Idea

My border idea would be simple. Since the Rio Grande River already exists between the southern border of the United States (Texas) and Mexico why not extend that river all the way from the Gulf of Mexico to the Pacific Ocean? This could be a joint effort between Mexico and the United States and think of the jobs it would open up.

Project Duration

Think about the amount of manpower, equipment, and resources needing to be provided for a massive workforce.

New Water Source

Think about the water source you would bring to land that has never seen water before. Could industry be brought to that region to convert salt water to fresh?

Think of the long-term farming possibilities if that was possible.

Waterfront Property

What about beach front or resort property? Where it would become available and may never have been possible. Tax dollars.

Curtailing Crime

In the US we always hear about the drug cartels and the tunnels they have made to cross under the borders of Mexico and the US. Dredging a deep river would expose and drown out all of those tunnels for drug trafficking.

Our New Border

The United States already has a working Navy and Coast Guard that can be useful alongside our current border patrols. The difference would be moving to boats through our new found waterway.

A Far-Fetched Idea

It would seem to me that if everyone in Mexico wanted to be in the United States of America, if they cannot come to America then America could come to them. What does that mean exactly? Well, many countries and or continents were taken over by other populations of other countries. Why spend all the money and labor to build a wall between the United States and Mexico or a waterway like I proposed in a previous chapter of this writing. Why cannot the United States get the Mexican population on their side, and the United States stake a flag in the country of Mexico and claim it in the name of the people of Mexico.

Seems simple enough. If the Mexican population rises up against their political figures than who would stop the transition of power? Think of it, Mexico the 51st state and the District of Columbia.

2 ... B O R D E R   C R I M E S

The U.S.-Mexico border is a major focal point for various types of criminal activity due to the large volume of people, goods, and traffic that pass through it daily. Law enforcement agencies such as U.S. Customs and Border Protection (CBP), the Drug Enforcement Administration (DEA), U.S. Immigration and Customs Enforcement (ICE), and others work together to address a range of crimes that can be intercepted or prevented along the border. Here are some of the key types of crimes that can be, and are, intervened through the U.S.-Mexico border:

D r u g   T r a f f i c k i n g

- Drugs such as cocaine, heroin, methamphetamines, fentanyl, and marijuana are frequently smuggled across the border. Criminal organizations, including Mexican cartels, use various tactics to move illegal drugs, including tunnels, hidden compartments in vehicles, drones, and even legitimate shipping containers.
- Law enforcement agencies, such as the DEA and CBP, work to detect and intercept drug shipments, conduct surveillance, and dismantle smuggling operations.

H u m a n   T r a f f i c k i n g   a n d   S m u g g l i n g

- Human trafficking involves the exploitation of individuals for forced labor, sexual exploitation, or involuntary servitude. This is a serious issue along the U.S.-Mexico border, as human traffickers often exploit migrants, particularly those from Central America, to transport them across the border for forced labor or sexual exploitation.

- Human smuggling involves the illegal transportation of individuals, often across the border, in exchange for payment. This is typically done with the cooperation of smugglers or criminal organizations.
- U.S. law enforcement works to identify and stop human trafficking rings and prevent illegal border crossings by smuggled migrants.

Illegal Immigration

- Unauthorized border crossings are a common issue. People seeking to enter the U.S. illegally (often for employment, asylum, or economic reasons) may attempt to cross the border without proper documentation.
- The U.S. Border Patrol is responsible for detecting and apprehending individuals attempting to cross the border illegally, often in remote areas or through areas of high traffic.

Weapons Smuggling

- Illegal firearms trafficking is another major concern. Criminal organizations on both sides of the border may smuggle firearms, often from the U.S. into Mexico, where they are used by drug cartels or other violent groups.

- ATF (Bureau of Alcohol, Tobacco, Firearms, and Explosives) and CBP work together to prevent the illegal flow of weapons and track illicit gun sales and trafficking routes.

Terrorism and National Security Threats

- While relatively rare, the U.S.-Mexico border is a point of concern for national security. Criminal groups, such as terrorist organizations, may attempt to infiltrate the U.S. or use the border to facilitate illegal activities. The U.S. Department of Homeland Security (DHS), along with Border Patrol, regularly monitors for potential security threats.
- Terrorist groups might exploit weak spots in the border security infrastructure to smuggle weapons, materials, or operatives into the U.S.

Money Laundering

- Money laundering operations often use the U.S.-Mexico border as a route for the transfer of illicitly gained funds. Criminal organizations may launder money through a variety of means, including under-the-table businesses, casinos, and through cross-border transactions.
- Agencies like the U.S. Treasury's Financial Crimes Enforcement Network (FinCEN) and ICE's Homeland Security Investigations (HSI) are involved in preventing and investigating money laundering across the border.

Organized Crime and

Cartels

- Mexican drug cartels and other criminal organizations control much of the illegal cross-border trafficking, including drug and human smuggling, extortion, kidnapping, and other violent crimes.
- Cartels and gangs involved in organized crime often use the border as a hub for coordinating illicit activities, such as controlling trafficking routes and conducting violent acts to protect their operations.
- FBI, DEA, CBP, and Homeland Security Investigations are involved in targeting these cartels and their operations.

Vehicle and Cargo Smuggling

- Vehicle smuggling involves concealing illicit goods, such as drugs, weapons, or people, inside cars, trucks, or even commercial cargo.
- CBP uses advanced technology, including X-ray scanners, to inspect vehicles and cargo at border checkpoints to detect hidden contraband and intercept smuggling operations.

Corruption and Border Security Breaches

- Corruption among border personnel (e.g., Border Patrol, customs agents) can lead to breaches in border security, allowing smuggling or unauthorized crossings. Investigating and

preventing corruption within the border enforcement agencies is a priority.
- ICE's Office of Professional Responsibility (OPR) and the CBP Office of Internal Affairs are responsible for investigating such cases.

Environmental Crimes

- Environmental violations along the border may involve illegal deforestation, waste dumping, or destruction of protected lands. Environmental groups have raised concerns over activities like illegal construction and mining in border areas that could harm the ecosystem.
- Agencies like the Environmental Protection Agency (EPA) and U.S. Fish and Wildlife Service (FWS) sometimes collaborate with law enforcement to address these issues in border regions.

Counterfeit Goods

- The U.S. and Mexico share a significant trade relationship, and illegal shipments of counterfeit goods (e.g., pirated electronics, clothing, pharmaceuticals) are sometimes smuggled across the border to evade customs duties or intellectual property laws.
- CBP enforces intellectual property rights (IPR) at the border to prevent the importation of counterfeit or pirated goods.

In Conclusion

To address these potential crimes, a combination of border security technologies, international cooperation, law enforcement agencies, and intelligence sharing needs to be utilized. While significant progress has been made in tackling these issues, the U.S.-Mexico border remains a hotspot for illegal activities, requiring continuous efforts to enhance security and enforce the law.

3 ... B E N E F I T S   O F   T H I S
W A T E R W A Y

Constructing a waterway, such as a canal or water transport route, between the United States and Mexico could offer a range of economic, logistical, environmental, and political benefits for both countries. While such a project would be a monumental undertaking, there are several potential advantages to developing a shared waterway between the two nations.

Here are the main benefits for both countries:

E n h a n c e d   T r a d e   a n d   E c o n o m i c   G r o w t h

- **Increased Trade Volume**: A new waterway could boost trade between the U.S. and Mexico by providing a more efficient, cost-effective route for shipping goods. This could especially benefit industries reliant on the movement of bulk goods (e.g., oil, agricultural products, manufactured goods).
- **Improved Access to Ports**: The U.S. and Mexico could both benefit from better access to each other's ports. For example, if the canal connected the Gulf of Mexico to the Pacific Ocean, it could facilitate easier passage for goods destined for either market.
- **Job Creation**: The construction of the waterway itself would create thousands of jobs in the short term. In the long term, the operation and maintenance of the waterway would also

generate jobs in transportation, logistics, and various related sectors.

- **Boost to Regional Economies**: Areas along the proposed waterway—especially ports, logistics hubs, and industrial zones—could see significant economic benefits, creating new opportunities for local businesses, as well as increased tax revenues for municipalities and states.

Strategic Geopolitical Advantage

- **Strengthening Bilateral Relations**: A major infrastructure project like this could foster closer cooperation between the U.S. and Mexico, strengthening diplomatic and economic ties. It would require extensive collaboration, fostering mutual interests and goals between the two countries.
- **Diversifying Shipping Routes**: A shared waterway could provide a strategic alternative to traditional shipping routes (e.g., the Panama Canal), enhancing both countries' positions in global trade. This would allow for more control over maritime traffic, especially in the event of geopolitical tensions or disruptions in other parts of the world.

Improved Transportation Efficiency

- **Reduced Shipping Times and Costs**: A new waterway connecting the U.S. and Mexico would offer more direct access between the two countries, cutting down on the time and cost of transporting goods. For example, a canal could reduce the need for long, expensive trips around the southern tip of South America or the need to navigate congested sea routes in other parts of the world.

- **Increased Capacity for Freight**: A waterway could be designed to handle larger vessels and higher volumes of freight compared to overland transport, which is often subject to congestion, weather, and infrastructure limitations.

Environmental Benefits

- **Reduced Road Traffic**: By shifting a significant portion of goods transportation from trucks to ships, a waterway could reduce road congestion, lowering carbon emissions and wear on highways. This would be a more environmentally sustainable option compared to increasing reliance on truck or rail transport.
- **Green Energy Integration**: If the waterway is integrated with renewable energy sources (such as wind or solar power for canal operations or electric vessels), it could provide an environmentally friendly alternative to traditional shipping methods.
- **Enhanced Water Management**: In some cases, waterways can be used for dual purposes, such as improving water supply and irrigation for surrounding areas. If such a project is integrated with Mexico's agricultural regions, it could benefit farming and water distribution in the region.

Tourism and Cultural Exchange

- **Tourism Development**: A new waterway could become an iconic project that attracts tourists from around the world, similar to the Panama Canal. Ports, ships, and surrounding areas could develop tourist attractions, offering cultural and historical tours, eco-tourism, and recreation, all of which would benefit local economies.
- **Cultural and Social Connectivity**: A shared waterway could increase people-to-people connections between the U.S. and Mexico, facilitating greater cultural exchange, tourism, and cross-border travel. This would help bridge the two countries' societies, fostering a stronger sense of community and shared interests.

Environmental and Resource Management

- **Waterway for Irrigation and Agriculture**: In addition to facilitating shipping, a canal could serve as a source of water for agricultural purposes, potentially benefiting regions in both the U.S. and Mexico that rely on irrigation. In Mexico, particularly in regions like Baja California and Sonora, access to water for irrigation can be a critical factor in agricultural productivity.
- **Improved Flood Control and Water Storage**: A canal system could be engineered to help with flood management and water storage, which

could be valuable for mitigating seasonal floods, especially in agricultural areas along the border.

Promoting Innovation in Engineering and Infrastructure

- **Technological Advancements**: The construction and operation of a major waterway would stimulate advancements in engineering, infrastructure technology, and transportation systems. It could spur innovation in the design of locks, vessels, cargo handling, and port facilities.
- **Global Positioning for Both Countries**: Successfully completing such an ambitious project would elevate the international standing of both the U.S. and Mexico as leaders in infrastructure development. It could position both countries as key players in global shipping and logistics.

Security and Border Control

- **Enhanced Border Control**: A shared waterway could allow for more streamlined and effective monitoring of cross-border traffic, improving the ability to track and intercept illicit goods (like drugs or weapons) or prevent illegal immigration. With modern surveillance and security systems in place, it could become a controlled, regulated corridor that benefits both countries' national security.

- **Collaboration on Border Enforcement**: With a joint waterway project, the U.S. and Mexico could work together on enforcement mechanisms, making it easier to monitor and ensure that goods and people moving across the border comply with legal standards.

In Conclusion

A waterway between the United States and Mexico could provide substantial economic, logistical, and political benefits for both nations. It could facilitate trade, create jobs, enhance regional cooperation, reduce environmental impacts from overland transport, and provide long-term strategic advantages. However, careful planning and collaboration would be essential to overcome the challenges involved in such an ambitious infrastructure project.

4 ... CHALLENGES OF THIS WATERWAY

While the construction of a waterway between the United States and Mexico could provide many potential benefits, there are also significant challenges that both countries would need to consider. Below are some of the key issues that could arise from such a massive project:

Environmental Impact

- **Ecosystem Disruption**: Building a waterway could have profound negative effects on local ecosystems. The construction process could lead to habitat destruction for wildlife, particularly in delicate areas like wetlands, forests, or rivers that are home to unique species. Altering waterways and landscapes could disrupt plant and animal life, causing long-term ecological damage.
- **Water Quality and Pollution**: A large waterway could lead to pollution from shipping traffic, including oil spills, chemical runoff, and other contaminants. Additionally, dredging and altering the course of rivers or coastlines could lead to changes in water quality, potentially harming local agricultural areas or affecting nearby communities' water supply.
- **Water Usage and Supply**: In areas where water resources are already scarce, the creation of a large waterway could lead to water diversion, affecting local agriculture, drinking water supply, and natural water systems. This could create conflicts over water rights,

especially in regions already dealing with water scarcity (e.g., Northern Mexico and Southern U.S. states).

Economic and Financial Burdens

- **High Construction Costs**: The cost of constructing a waterway of this scale would be enormous, potentially in the **tens of billions of dollars**. Funding such a project could strain national budgets, lead to higher taxes, or divert funds from other essential projects (e.g., healthcare, education, infrastructure development). Financing the project may involve debt or long-term loans, which could burden future generations.
- **Operational and Maintenance Costs**: Beyond construction, maintaining the waterway — including dredging, security, infrastructure upkeep, and operational costs — would be a continuous financial commitment. The cost of keeping the canal operational and safe from issues like blockages, accidents, or infrastructure failure could place a long-term burden on both governments.
- **Economic Displacement**: While some jobs would be created, others could be displaced. Industries reliant on other transportation modes (like trucking or railways) may suffer from the reduction in demand for land-based shipping. Workers in these sectors may face unemployment or the need for retraining. Additionally, certain local economies that depend on other industries (like agriculture or

tourism) might face challenges if the waterway disrupts regional economies.

Geopolitical and Security Risks

- **Security Vulnerabilities**: The creation of a waterway could open up new security challenges. A major shipping route would be an attractive target for smuggling, piracy, or even terrorist activities. Criminal organizations (like drug cartels) could exploit the waterway for trafficking narcotics, weapons, or people, especially if security infrastructure is not robust enough to handle the increased traffic.
- **International Tensions**: While the U.S. and Mexico could collaborate on a waterway, disputes over the management, regulation, or control of the waterway could arise. For example, there may be disagreements over who controls access, shipping rights, tolls, or environmental regulations. If there are conflicting interests, it could strain diplomatic relations.
- **Impact on Sovereignty**: A shared waterway may lead to political and legal challenges over the governance of the project, with both nations having competing priorities. Joint control could result in bureaucratic inefficiency, delays, and disputes over how to manage security, trade policies, or environmental concerns.

Social and Cultural Disruption

- **Displacement of Communities**: Constructing a canal or large waterway often requires significant land acquisition, which could lead to the displacement of local communities. People who live along the border or in areas that would be affected by the canal may have to relocate. This could cause social upheaval, loss of livelihoods, and resentment toward the project.
- **Cultural and Heritage Concerns**: Many areas along the U.S.-Mexico border have rich cultural and historical significance. The construction of a large waterway could endanger heritage sites, sacred lands, or indigenous territories. Local communities may resist the project if they feel their cultural or historical heritage is being ignored or harmed.

Impact on Agriculture and Local Industries

- **Agricultural Land Loss**: The land required to build a waterway could displace existing agricultural operations, particularly in regions like Northern Mexico and the southwestern United States. Large tracts of fertile farmland might be lost, affecting local food production and the livelihoods of farmers.
- **Disruption of Trade Routes**: While a waterway could boost trade in some sectors, it might also disrupt established trade routes and supply chains. Regions that depend on rail or truck transport might see increased competition, and

businesses involved in those industries may face financial difficulties.

Environmental and Climate Change Concerns

- **Changing Water Systems**: Construction of a large waterway could alter natural water systems, potentially changing local hydrology. If the waterway were to redirect water or interfere with river flow, it could exacerbate flooding, drought conditions, or soil erosion in surrounding areas, particularly in a region already vulnerable to such issues.
- **Climate Change Vulnerability**: The waterway infrastructure, particularly if it's near coastal areas, could be vulnerable to rising sea levels and other climate change impacts. The canal and its surrounding infrastructure could face increased risks of flooding, damage from storms, or long-term environmental degradation, especially if climate change accelerates in the coming decades.

Political and Legal Challenges

- **Complex Legal Framework**: The construction of a waterway that crosses international borders would require extensive legal agreements between the U.S. and Mexico. This would involve intricate negotiations on everything from land rights and water usage to security and environmental protection. Disputes over these issues could delay construction or lead to legal challenges, particularly if there are disagreements over governance or jurisdiction.

- **Regulatory and Environmental Compliance**: Both countries would need to navigate complex regulations to ensure that the waterway meets international environmental standards, navigational laws, and maritime treaties. Disputes over regulations could lead to delays, litigation, or environmental harm.

Technological and Engineering Risks

- **Construction and Design Challenges**: Building a canal or large-scale waterway through difficult terrain (desert, mountains, or forests) presents enormous technical and engineering challenges. The risk of delays, cost overruns, and engineering failures would be significant, particularly in the face of unexpected geological or environmental conditions.
- **Technological Obsolescence**: As technology in shipping and logistics continues to evolve, the demand for a massive, fixed waterway might diminish over time, particularly if alternative transportation technologies (like automated shipping, drones, or high-speed rail) become more efficient. This could make the project less relevant or economically viable in the long term.

In Conclusion

While the idea of building a waterway between the U.S. and Mexico holds many potential benefits, the detriments are significant and must be carefully weighed. The project would face enormous financial costs, environmental risks, geopolitical tensions, and social challenges. It would also require careful management to mitigate security threats, economic disruptions, and cultural or legal disputes. Both governments would need to conduct thorough feasibility studies, environmental impact assessments, and public consultations to determine whether such a large-scale infrastructure project is genuinely in the best interest of both countries, especially in the long run.

5 ... THE CURRENT UNITED STATES BORDER

The border between the United States and Mexico is approximately 1,954 miles (3,145 kilometers) long. It stretches from the Pacific Ocean in the west, through deserts and mountains, to the Gulf of Mexico in the east.

Approximately 120 miles (193 kilometers) of the U.S.-Mexico border is made up of water. This includes the sections along the Rio Grande and other rivers, as well as the Gulf of Mexico coastline.

The Rio Grande forms a significant portion of the border in the western part of Texas, and there are additional smaller bodies of water along the border in places like the Colorado River and its tributaries. The border's water sections also include the mouth of the Rio Grande where it meets the Gulf of Mexico.

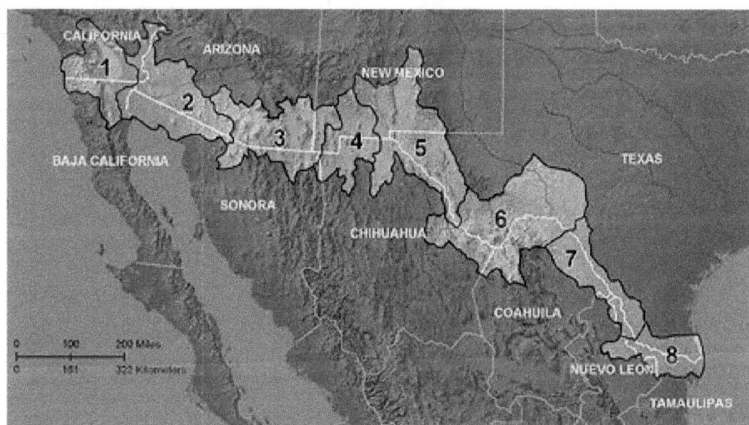

Four U.S. states share a border with Mexico:

- California
- Arizona
- New Mexico
- Texas

These states have varying lengths of border with Mexico, with Texas having the longest and California the shortest. The border runs from the Pacific Ocean in the west (California) to the Gulf of Mexico in the east (Texas).

The U.S.-Mexico border stretches across four states and includes multiple counties within each of those states. Here's a breakdown:

California

- **Counties**:
  - **San Diego County**

California's border with Mexico is the shortest, stretching about 140 miles from the Pacific Ocean to the eastern border of San Diego County.

Arizona

- **Counties**:
  - **Yuma County**
  - **Pima County**
  - **Santa Cruz County**
  - **Cochise County**

Arizona shares a border with Mexico that is about 370 miles long. The border crosses through deserts, mountains, and stretches of land near major cities like Tucson.

New Mexico

- **Counties**:
    - **Hidalgo County**
    - **Luna County**
    - **Doña Ana County**

New Mexico's border with Mexico is around 180 miles. The region includes areas such as the southern parts of the state, including the city of Las Cruces.

Texas

- **Counties**:
    - **El Paso County**
    - **Hudspeth County**
    - **Culberson County**
    - **Jeff Davis County**
    - **Brewster County**
    - **Pecos County**
    - **Val Verde County**
    - **Kinney County**
    - **Maverick County**
    - **Zapata County**
    - **Webb County**
    - **Dimmit County**
    - **La Salle County**
    - **McMullen County**
    - **Jim Hogg County**
    - **Starr County**

- Hidalgo County
- Cameron County

Texas has the longest stretch of border with Mexico, about 1,254 miles. Texas shares a border with Mexico from the western part of the state near El Paso, all the way to the Gulf of Mexico. The counties along the Texas-Mexico border range from urban areas like El Paso and Laredo to more rural and agricultural regions.

In Conclusion

This vast border area includes a diverse mix of landscapes, from urban regions to deserts, rivers (like the Rio Grande), and agricultural land.

6 ... THE CURRENT MEXICO BORDER

Mexico shares a border with the United States through six states, and each of those states has various municipalities (equivalent to U.S. counties) that lie along the border.

Here's the breakdown:

Baja California

- **Municipalities**:
  - **Tijuana**
  - **Mexicali**
  - **Ensenada**

Baja California is the northernmost state of Mexico and borders California. The city of Tijuana is the largest and most prominent city along the U.S.-Mexico border in this state, while Mexicali is the capital.

Sonora

- **Municipalities**:
  - **San Luis Río Colorado**
  - **Caborca**
  - **Nogales**
  - **Hermosillo** (some parts)

Sonora borders Arizona and features cities like Nogales and San Luis Río Colorado, which are directly adjacent to U.S. border cities of the same name.

Chihuahua

- **Municipalities**:
  - **Juárez**
  - **Palomas**
  - **Ojinaga**

Chihuahua borders Texas and is home to Ciudad Juárez, one of the largest cities on the U.S.-Mexico border, across from El Paso, Texas. It is one of the most significant border crossings.

Coahuila

- **Municipalities**:
  - **Ciudad Acuña**
  - **Piedras Negras**

Coahuila borders Texas and includes Ciudad Acuña and Piedras Negras as key border cities.

Nuevo León

- **Municipalities**:
  - Nuevo Laredo

Nuevo León borders Texas, and its major city is Nuevo Laredo, which is directly across from Laredo, Texas. This is one of the busiest border crossings in the world.

Tamaulipas

- **Municipalities**:
  - Reynosa
  - Matamoros
  - Nuevo Laredo

Tamaulipas borders Texas and has cities like Reynosa and Matamoros, which are crucial points for cross-border trade and travel.

In Conclusion

These states and municipalities make up Mexico's border with the United States, which spans a variety of environments, from bustling urban areas like Tijuana and Ciudad Juárez to more remote and desert-like regions. The border cities are key to both U.S.-Mexico trade and the cultural exchange between the two nations.

7 ... A SMALLER HISTORIC UNDERTAKING: THE PANAMA CANAL

The Panama Canal is approximately 50 miles (80 kilometers) long. It connects the Atlantic Ocean (via the Caribbean Sea) to the Pacific Ocean, cutting through the Isthmus of Panama in Central America. The canal is a vital waterway for global maritime trade, allowing ships to avoid the lengthy and dangerous route around the southern tip of South America.

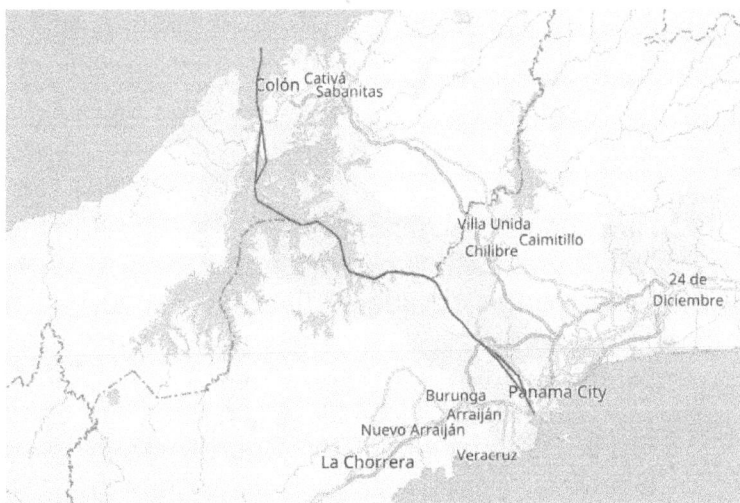

The U.S. president during the construction of the Panama Canal was Theodore Roosevelt.

Roosevelt played a crucial role in the building of the canal, both in terms of diplomacy and support for the project. In 1903, during his presidency, the U.S.

negotiated a treaty with Panama (after Panama gained independence from Colombia) to secure control of the canal zone. The construction of the canal began shortly after, in 1904, and it was completed in 1914, several years after Roosevelt left office.

Roosevelt's leadership in securing the canal's construction and the subsequent U.S. control of the zone was a defining aspect of his foreign policy, demonstrating his belief in a strong U.S. presence in the Western Hemisphere, which he famously encapsulated with the phrase, "Speak softly and carry a big stick."

The construction of the Panama Canal took 10 years to complete, from 1904 to 1914.

Here's a breakdown of the timeline after the initial French attempt was abandoned:

- American construction (1904–1914): After the United States took control of the project in 1904, following the signing of the Hay-Bunau-Varilla Treaty with Panama, the construction of the canal began in earnest under U.S. leadership. The Americans, led by engineers like John Frank Stevens and later George Washington Goethals, focused on overcoming the challenges of disease control (with the help of Dr. William Gorgas, who helped eliminate mosquitoes) and the immense engineering obstacles.

The canal was officially completed and opened on August 15, 1914, and it became a vital global shipping route, drastically reducing the time and distance required for ships to travel between the Atlantic and Pacific Oceans.

The construction phase lasted 10 years under American control.

The construction of the Panama Canal required a massive amount of manpower, particularly during the American phase of construction (1904–1914). The total number of workers involved in the project varied over time, but here's a general breakdown:

American Phase (1904–1914)

- The U.S. effort to complete the canal was much more labor-intensive. At its peak, the American workforce included around 40,000 to 45,000 workers.
- The workforce was composed of both skilled and unskilled laborers, including engineers, supervisors, doctors, and manual laborers.
- The labor force was highly diverse, consisting of:
  - Caribbean laborers, mainly from Barbados, Jamaica, and other West Indies. They were hired for manual labor, particularly for digging, shoveling, and other physically demanding tasks. Around 20,000 workers were from the Caribbean at one point.
  - American workers: Skilled labor, engineers, and administrative personnel. These workers were generally better

compensated but represented a much smaller portion of the workforce.

○ Panamanian and other Central American workers: As the construction progressed, many local Panamanians and others from Central America also contributed to the labor force.

Total Manpower

- In total, over the course of the project, around 75,000 to 80,000 people worked on the Panama Canal, though many of these were laborers who worked for only short periods due to the harsh conditions.
- The U.S. project also had a significant emphasis on disease control. Under the leadership of Dr. William Gorgas, a concerted effort was made to eliminate mosquitoes that carried yellow fever and malaria, which helped reduce the death rate among workers and allowed the project to move forward successfully.

Worker Conditions and Casualties

- The work was grueling and dangerous, with a high mortality rate in both the French and American phases. During the French effort, thousands of workers died due to disease and accidents.
- The American effort saw a marked improvement in health conditions due to better sanitation and mosquito control, though there were still many

worker deaths, particularly among the Caribbean laborers.

In Conclusion

In sum, the Panama Canal's construction involved tens of thousands of workers, many of whom faced perilous conditions, but their collective labor was key to completing one of the most ambitious engineering feats in history.

8 ... THINKING LEADING TO IDEAS

Thinking leads to the coming up with ideas. The ideas we come up with are presented and made available to others to review. In this chapter I am going to talk about thinking, the three types of thinking that have enabled me to come up with the idea I have presented in this writing: critical, creative and strategic.

Many years ago the thinking was right-brain and left-brain. It was though that human elements had one or the other as dominant. You were either a creative or critical thinker, not 100% but more of one than the other. It sounds logical, right? Some of us are better at being creative and some of us are hard and deep thinkers, right? Then some of us think we are just the "cat's meow" at everything. My question would be: What does everyone else think about that?

I like to think of creative and critical thinking as working together. Kind of like a "Yin and Yang". Both things make you complete. I like to think of them as two brothers working together to solve a common goal. Then we have this thinking method called strategic thinking. Strategic thinking is like the cousin of two brothers. You know the third wheel when you get together, the guy who can't make a decision without looking at both sides of all issues, you know the politically correct thinker that "beats the dead horse" so-to-speak. You don't mind him being around or coming to the party, you just want him to get to the point!

Thinking is a series of processes we do not realize, as happening every second of every minute of every day. Pulses bouncing from left to right and right to left in our every waking moment as well as when we are at rest. Some thoughts we control, or try to control, and some thoughts we may not even recognize are being processed through our subconscious, while sleeping.

The human mind is always trying to *work things out*, to *decipher*, to *interpret*, and to *untangle* pieces of data the physical senses bring to its attention: sight, smell, hearing, taste, and touch. The senses open up a door of awareness to our minds. Once the door is opened our mind starts to analyse and process what just happened, and we usually don't even realize it is happening, it happens sometimes in the background. This is the *blank slate* each human is born with that starts to compile information form the first day it is brought into the light and continues until the final day when that light has extinguished.

During the lifecycle of this *blank slate*, foreseen and unforeseen issues and problems abound over the years. Issues that can be small or large, and simple or complex. Issues that are either under our control and or not of our making but that require our input, for a solution. When a problem or issue is sensed the brain starts to go through analysing and processing the received data.

A solution for this sensed problem or issue is limited by the current knowledge base of the recipient, and in fact each human entity is a product of its environment and we all have limitations of previous experiences and familiarities to help with our resolve.

There are two sides to the human brain, the left and the right. Both sides work together sending pulses back and forth. Some people are left-brain dominant, some are right-brain dominant and some use both-sides somewhat equally, which is a lesser percentage than the left and right dominant. People who are labelled as left-brain thinkers are said to be logically and analytically dominant. People who are labelled as right-brain thinkers are said to be innovatively and creatively dominant.

I think you get the point, enough about that, let's move on. The proceeding pages of this chapter are relative to the left-brain dominant, logical *critical* thinker and the right-brain dominant, innovative *creative* thinker and the thinker that combines the two, the strategic thinker, maybe even doing this unknowingly. Problems and issues, are sensed and analysed by the brain where it decides what tools are needed to help resolve the situation. Remember now, each human has a limited knowledge base, limited by the experiences of what they have learned and acquired from each of their respective environments. Sometimes we need to use our logical side for resolve and sometimes we need use our creative side. Some of us are better at some things and others other things. Sometimes we need to think logically for a solution and sometimes we need to look at being creative.

Critical Thinking

When I think of critical thinking, I think of logic or thinking logically. That may be a stretch to others in the field, but that is how I see it. I see it as looking at an issue or problem and realistically looking for answers to solve it, as your brain scans through the folders of knowledge you have attained throughout your short existence to find a solution. What process does the human mind need to go through to solve that issue or problem? Is it a small process with only a few steps, or is it a larger problem that requires many steps and/or sub-steps within the steps?

As with the critical thinker, there are many ways to articulate the concept of their thinking, critically that is. Yet every substantive conception of critical thinking must contain certain core elements. Consider the following brief conceptualizations.

"Critical thinking is the intellectually disciplined process of actively and skillfully conceptualizing, applying, analyzing, synthesizing, and/or evaluating information gathered from, or generated by, observation, experience, reflection, reasoning, or communication, as a guide to belief and action. In its exemplary form, it is based on universal intellectual values that transcend subject matter divisions: clarity, accuracy, precision, consistency, relevance, sound evidence, good reasons, depth, breadth, and fairness..."

"Critical thinking is self-guided, self-disciplined thinking which attempts to reason at the highest level of quality in a fair-minded way. People who think critically consistently attempt to live rationally, reasonably and empathically. They are keenly aware of the inherently flawed nature of human thinking when left unchecked. They strive to diminish the power of their egocentric and socio-centric tendencies. They use the intellectual tools that critical thinking offers – concepts and principles that enable them to analyze, assess, and improve thinking. They work diligently to develop the intellectual virtues of intellectual integrity, intellectual humility, intellectual civility, intellectual empathy, intellectual sense of justice and confidence in reason. They realize that no matter how skilled they are as thinkers, they can always improve their reasoning abilities and they will at times fall prey to mistakes in reasoning, human irrationality, prejudices, biases, distortions, uncritically accepted social rules and taboos, self-interest, and a vested interest."

"Critical thinkers strive to improve the world in whatever ways they can and contribute to a more rational, civilized society. At the same time, they recognize the complexities often inherent in doing so. They strive never to think simplistically about complicated issues and always to consider the rights and needs of relevant others. They recognize the complexities in developing as thinkers, and commit themselves to life-long practice toward self-improvement. They embody the Socratic principle: *The unexamined life is not worth living*, because they realize that many unexamined lives together result in an uncritical, unjust, dangerous world."

Within the industrial working environment, this can be as simple as working on an assembly line and putting nut 'A' onto bolt 'B'. How many steps do you need to draw or spell out to do that simple job or task? Can just anyone or everyone do that simple task? Does everyone need the same number of steps to complete that task? Does the human mind work in all matters the same for everyone? Think about it. Do you think the same way that everyone else around you does? That is doubtful, as if you know anything about the human mind, you know that everyone is different. If all the writings throughout time are right, then each human entity or in relation to the title of this reading, the college freshman, is a product of their environment and surroundings. If correct, these are not the same for everyone. Every potential student is different.

Then there comes the additional factor of age: the age of the college freshman. If we look at statistics in relation to the college freshman, the majority of that population have just graduated high school or have graduated a few years earlier, generally a population between the ages of eighteen to twenty-two. The older we are and the more time we have spent breathing and accessing the realizations throughout our lives, the more blocks of knowledge we have acquired and retained. Our minds retain those ideas, concepts, and solutions, whether we realize it or not. From birth, the human mind is like a sponge soaking up every bit of knowledge your physical senses pick up. Some human senses are stronger and keener than others. After all, we are all different in our makeup, Right? A newborn child has often been spoken of as a *clean slate*: a *clean slate* that progressively soaks up everything it hears, sees, and feels, leading through the years to what we are as we grow older.

Let's look at a more intensive/extensive example of necessary steps to achieve a solution. What if you are thinking of an education and the educational pathway is something within the realm or area of computer science and/or information technology? I choose these realms as these are what I have been involved in throughout my existence on this planet. What issues or problems might a computer science or information technology student run into during their pathway to educational success in the classroom?

Let's first look at computer science. First off, computer science is one of the most difficult post-secondary educational programs that exist today; it is likened to engineering and the sciences. This is not to say they are the only difficult educational pathways; these are the ones I am familiar with and have knowledge of. This is not to deter you, but rather to prepare you. The career positions tied to these programs of study can be highly lucrative and pay very well. That is generally why a lot of potential students seek out these programs. If we are seeking out these programs, we need to understand what is required to be successful in them. After all, we all want to be successful, right? The purpose of getting an education is to graduate and get a good career, right? And of course, make good money, right? Why not reap the benefits that come with it, right?

Within computer science, there are a lot of programming courses, and some are statistics-related. These courses generally take more preparation and mindful thinking to process: to process and successfully work through the problems you may be given in your coursework to find correct and efficient solutions. These problems will not be as simple to solve as the nut 'A' on bolt 'B' example mentioned earlier. These will be more complex with multiple steps and sub-steps. Pages of programming languages and code to work through, to troubleshoot, and to develop.

How about information technology? When you think of information technology, you think of computers, hardware and software for computers, the networking of computers, and the security of all these things. There is a lot of memorization of the concepts and ideas within this coursework, as well as steps and sub-steps of troubleshooting devices and the planning and design of computing networks. Additionally, you will also be involved in programming related to computers and networked devices as well as website languages. These can be somewhat complicated, but they are all guided by the student and their desire to achieve and succeed. The object is not to dissuade you from seeking these programs but to inform and prepare you. You will surely come to use your Ten Gallon Hat. It will become your second nature.

C r e a t i v e   T h i n k i n g

Some of us are more 'creative' than others. Some of us are better artists than others. Some of us design, draw, paint, and illustrate, and some of us just don't seem to have that talent. Can you acquire that talent, or is it something you are either 'born with' or not? Well, in previous wording, we mentioned that human entities and authors throughout time have said babies are thought to be born with a *clean slate*, right? That is what has been said. If that is true, then that 'baby' must be trainable or teachable in some form or fashion, right? If the human entities of the world are correct, then it must be.

You just have to come into those situations where your mind needs to view what it sees in a different manner.

Looking at an issue or problem through a different *lens* than you normally see. Have you ever heard someone say "they are looking at that with rose-colored glasses?" What does that mean? That means that you are looking at something with bias. That means you are looking at something without an open mind; you are only seeing what you want to see. You are not looking past your own knowledge base. You are only trying to solve that issue with the knowledge you have attained so far through your life and not looking past that, as you may not know everything and there may be more knowledge you can attain past the point of today, at any given moment in time. Does that make sense?

The creative thinker continually digs deep into oneself and generates more, newer, better, faster, cheaper, different ideas that can be used to improve the important parts of their life, i.e. the successful manager. The creative thinker is comprised of seven distinct qualities:

1. I am curious to a fault
2. I practice zero-based thinking
3. I am willing to change
4. I am goal focused
5. I am willing to admit when I am wrong

6. I do not know everything
7. My ego will not be bruised if and when I am proven wrong

The first quality of creative thinkers is that they are curious to a fault. They are always asking questions. They never stop asking questions, and would like to find others similar to themselves. They ask questions like "Why or Why not, why can't we do that, if it hasn't been done, can we do it now?" There onslaught of questioning would be similar to that of a child, but they know why they ask, they have reasons, other than the bantering of a child.

The second quality of creative thinkers is that they practice "Zero-Based Thinking" throughout their daily routines. They continually ask themselves, "If I were not now doing what I am doing, knowing what I now know, would I start?" If their answer is no, they stop that train of curiosity , then move on to another thought project and start their questioning all over again. They say that hindsight is always 20-20. Creative thinkers move forward and do not dwell on the "what if" scenario. So many humans persist on questioning and spending vast amounts of energy on projects that when they look behind they would have never started to begin with, and generally wish they had not, then they wonder why they make so little progress and it seems to take forever.

The third quality of creative thinkers is that they are always willing to change, they are open to new suggestions and have an open mindset. They prefer to be in charge of their lives rather than being caught up in a flash flood of change that may often be inevitable and or unavoidable. The words of the truly flexible person, the person who is willing to change are simply, "I changed my mind." According to researchers, fully 70% of the decisions you make turn out to be wrong in the long run. This means that you must be willing to change your mind and try something else the majority of the time. Mental flexibility is the most important quality that you will need for success in this the 21st century.

The fourth quality of creative thinkers is their willingness to admit when they are wrong. They do not hold fast to their opinions when they are proven wrong. Researchers say that 80% of people burn up most of their mental and emotional energy defending against admitting that they made a wrong decision. True creative people are open minded, fluid, flexible and willing to both change their mind and admit that they are wrong when their earlier decisions turn out to be incorrect.

The fifth quality is that creative people can say, "I don't know." They recognize that it is impossible for anyone to know anything about everything, and it is very likely that almost everyone is wrong to some extent, no matter what they are doing. So when someone asks them a particular question that they don't know the answer so they admit it early and often. They simply say, "I don't know." And if necessary they go about finding the answer. Here's an important point. No matter what problems you have, there is someone somewhere who has had the same problem and who has already solved the problem and is using the solution today. One of the smartest and most creative things you can do is to find someone else, somewhere, who is already implementing the solution successfully and then copy him or her. The smartest person is not necessarily the person who comes up with the idea, but the one who accentuates or successfully builds upon it.

The sixth quality of creative people is that they are intensely goal focused. They know exactly what they want. They have it written down very clearly. They visualize it on a regular basis. They imagine what their goal would look like if it were a reality today. And the more they visualize and imagine their goal as a reality, the more creative they become and the faster they move toward achieving it.

The seventh quality of highly creative thinkers is that they have less ego involvement in being right. You will not bruise or crush their ego if you prove them wrong. They are always looking for and willing to accept the correct answer even if it down not come from them. They are more concerned with what is right rather than who is right. They are willing to accept ideas from any source to achieve a goal, overcome an obstacle, or solve a problem.

The most important part of creative thinking is the ability to generate ideas. And the greater the quantity of ideas that you generate, the greater the quality your ideas will be. Similar to that of a brain trust: a gathering of multiple brains coming up with multiple ideas to successfully come up with a solution. The more ideas you have, the more likely you are to have the right idea at the right time. But generating ideas is only 1% of the equation. As Thomas Edison once wrote, "Genius is one percent inspiration and 99 percent perspiration."

Your ability to come up with an idea, to test it and validate it, and then to implement it through creative thinking and execute it in your life to achieve results is the true mark of a successful IT manager. Every single time you originate a new idea, write it down, make a plan for its implementation through creative thinking and then take action, you are behaving like a genius. And the more you manage your creativity in this way, the smarter you will become. And the smarter you will become, the more you will achieve in every area of your life.

It is a rather drawn-out and lengthy concept. Basically, it means not to let the limits of what you have learned *soaked up* so far in life deter you from being able to *soak up* new things. After all, none of us know everything; we are only human entities and we are always learning, even when we grow older, we never stop learning. You may think you have stopped learning, but remember in previous paragraphs we talked about the human mind as *soaking up* as it goes along. Well, until we pass through this life, the thought is that we always gather information whether we realize it or not.

Enough explanation and background, let's get to the title at hand: Creative Thinking. Looking at what we need to do and the steps we may need to make to solve an issue or problem creatively. Could we call this thinking 'outside the box'? What does that mean? That means to find a way to solve the issue or problem at hand by using something other than the normal process and by using some mental tool or reasoning each of us may have attained throughout our lives, each one of us separately. Remember, when we talked earlier about the knowledge each of us attains throughout our daily lives and the environments we each are involved in? Remember that? That is what we are talking about. Since we all are a product of our environments since birth, maybe you have come into contact with something to help you personally solve the issue, problem, or task given. Then again, maybe not. Maybe you have not been privy to such an environment and have been somewhat sheltered or are quite young and have not experienced such situations in your limited past?

If I look at this through faculty eyes, we are often given the task of creatively finding a way to reach students. Past history shows educators that the more *senses* we can reach within a student, the more apt we are to give them an understanding of the material we are trying to convey. This brings to mind an example from my past when I was in an Associate Degree program in Computer Electronics. In one of the courses, we each bought a kit. A kit is a lot of pieces of electronic and electrical parts. These combined parts were an AC to DC voltage regulator power supply. We each bought a kit, and the object was to follow the directions and solder up and wire the kit. This kit would be used as our personal equipment in our future higher-level electronic courses. The day came when we were to plug in our power supplies for the first time to see our new inventions light up and get a passing grade.

Before I get to the *punch line*, I will give a little explanation. I transferred to this college from a Flight school in the south. At that Flight school, I was in a highly disciplined environment. At that college, I was one of the top students in my class and was often prodded by my instructors. All the instructors were ex-military, and all students towed the line. Since my father was a drill instructor in the Air Force, and that is how I grew up, I knew my place and my responsibilities. Well, if I wasn't in class at 7:00 am in the morning, the instructor would come beat on my dorm door until I answered. To make a long story short, this instructor made sure I was the best I could be.

The best I could be was to make mostly B's and some A grades for courses. My coursework was beaten into my head at that facility, and when I transferred to another college up north, I made all 'A' letter grades on everything! And I mean everything! Thanks to my previous instructors and their pushing me to be my best! At my new college, I was quickly made the course mentor and assistant for all my electronic/electric courses. I was the kid that helped the struggling students.

Back to the story and the punchline. I plugged my power supply in and passed the class without a hitch. My best friend plugged hers in and BANG! Smoke filled the room! It was like an old Batman serial rerun! Bang! Boom! Pow! But with the addition of a lot of smoke! And the smell was horrible! Have you ever smelled burning electronics? Well, needless to say, it wasn't pleasant, and it does have a smell all its own. Well, nobody got hurt as each power supply was in a metal case with knobs and leads. A good thing, right?

My friend was heartbroken. Something within her power supply exploded. She opened the case, cleared the smoke, and looked to see what had gone wrong. She could not find the problem as she looked through the *rose-colored glasses* we mentioned above. She could see the problem as she was looking at the issue with the bias that she had done everything correctly as per the instructions. She pushed it over to me and asked if I would take a look. In less than a minute, I saw the problem. I pushed it back to her and told her to really look! I asked her a few questions.

One was: Throughout ALL the classes we have had in this program talking about electrical and electronic devices, are there any of them that are directional? Meaning they can only go one way! She looked again. Aha, she was able to see it now, she put the capacitor in backward! In a capacitor, one lead is positive and one lead is negative. If you put it in backward, it will explode! Well, it exploded! Luckily it was inside a metal case as it was a rather large capacitor.

What does this story have to do with creative thinking? If I were the instructor of an electronic course and I purposely had students solder in a capacitor in the wrong direction, then the student energized the device. Trust me: That student would always remember the BANG and the burnt electronic smell for the rest of their life where they would not let it happen again. Would this not be a creative way of thinking to get a concept or idea into the mind of a student? I rest my case.

Through all my years of being in management and supervisory positions, giving presentations, and being the previous owner of an IT corporation, I have seen ideas from many authors discussing and laying out what they think are traits of managers and supervisors. In addition to being a manager or supervisor, which is a leader in many aspects, they also need to add traits of a follower as they are not the top tier of the corporate structure; everyone has a boss, right? Below I will propose what I think to be additional hats relative to leaders and followers, which I feel run parallel to the college freshman for their goal in achieving more than simple success and passing, but for attaining above-and-beyond success. The success that needs to be achieved by the graduating senior in their attempt to compete in the marketplace with all the other graduates around the world. The layout I propose is to provide individuals a planned method to *think* cohesively to be more effective in attaining results. What does that mean?

The foundation of this layout is that the human brain goes through a number of instructions or steps, if you will, to develop a plan of attack to solve a problem. Each instruction or step is required and is a necessity to solve the problem or task successfully. The thought is that each instruction or step when activated will unconsciously bring forth certain aspects of thought for the other instruction or step to be triggered, therefore coming up with a viable solution.

I will say this. These *instructions or steps* are not ways of thinking that all college freshmen share or may even possess, but I do think, however, that these *instructions or steps* can be a learned trait. These instructions or steps can be attained through educational reinforcement and practice. Growing up, didn't we always hear *Practice makes perfect*? Well, here we are, practice does make perfect, at least we are told. The more you practice, the better you get at it, right? Is it not easier to review work that you know has been successfully completed, as in a template, than it is to start from scratch with no provision of an example? It is like asking someone to reinvent the wheel when you have access to a perfectly good example of a wheel, right? What is the point?

Do we need an example here? Okay, let me set the stage. The first post-secondary school I went to for a degree was a Flight college in the south. I was an Avionics major, and my roommate was a Flight major. My roommate had some experience in flying and he wanted to get more flight time and a degree. He was very particular about his aircraft and didn't leave many things to chance when he flew. It seemed as though he was always flying when other students were just kind of hanging out. One of the things he liked was that one of the flight instructors had a similar background as he had, and he also wanted to get a lot of flight time, so they generally scheduled their flights together, instructor and student.

In the realm of flight, there are two ratings pilots achieve: VFR and IFR. VFR is Visual Flight Rating, and IFR is Instrument Flight Rating. Both of these ratings deal with eyesight. Visual is what you see inside and outside the cockpit and how you deal with what you actually see visually and how you react and deal with it. Instrument is where you ONLY use the instruments to see and project what to do next in a flight situation.

Make sense? So, my roommate had achieved both of these ratings. Remember, I said my roommate had spent a lot of time with one certain instructor during his time at the college. SO, on to my story.

I came back to the dorm one night, all the lights were out, and my roommate was sitting on the edge of his bed, very quiet. This was unlike him, so I knew something had happened that may be life-changing. Four hours ago, my roommate was up with a different instructor getting some extra flight time. He had to choose this other instructor as the one he usually scheduled with was up in the air with another student. My roommate and his new instructor are up in one airplane, and his usual instructor and another student are up in another. I don't remember what class-level storm came up, but it was rough, and both airplanes were caught in the storm. For over three hours, both of these airplanes were buzzing around the sky. Lightning struck my roommate's airplane, taking out all the internal and external lights. The only light still on were the instrument lights. The cockpit was dark, and you could not even see the one person sitting next to you.

The only time you could see something outside the windows was when the lightning flashed. The radio was damaged also, and every now and then, you could hear garbled talk from the other airplane and the flight base towers: In and out, in and out, bits and pieces of conversations. All my roommate could really hear was someone talking about 1000 feet. He didn't know whether it meant his aircraft was to go to 1000 feet or if the other airplane was to be at 1000 feet. He could not make out the whole conversation. The new instructor with my roommate was not at all comfortable with the situation, so my roommate took charge. Like I said, my roommate was a particular guy and had no problem taking over. He was confident in his training. If anyone knows anything about an airplane pilot, they are cocky and confident!

After several hours of struggle, it is pitch black outside, and lightning is flashing. My roommate hears an airplane engine, and it is close by. The lightning flashes, and right in front of my roommate's airplane is the other plane directly in front of him at the same altitude. My roommate said his whole life flashed before his eyes, and during the brief lightning strike, he could see the eyes of the others in the other airplane. He didn't hesitate, pulled up on the yoke, and he said it was like his aircraft was standing still, without motion. In that brief instant, it was over! All four were safe! The instructor of the other airplane pushed his yoke down.

The two students and instructors are alive today because my roommate was in charge of one airplane, and his usual instructor was in charge of the other. Those two had trained together many, many times for such a scenario as this.

My roommate is a highly successful pilot for a major airline today, and we talk all the time. Practice made this perfect.

How about another quick example, a little different? I tell you to run a race. I give you the starting point and the ending point, but I leave out any specifics or limitations of the race. Then YOUR mind begins to start sorting, calculating, and planning. Do I have enough information? Do I need to start at a certain time? Do I need to get to the endpoint at a certain time? What mode of transportation am I to use? How many other questions would or could a participant ask? We already can see how creativity ties in with the success of a student from the information previously presented. Creativity or Creative Thinking being one of the two main necessities mentioned in detail above along with Critical Thinking.

S t r a t e g i c   T h i n k i n g

The strategic thinker is an extension of the critical and creative thinker and is split into four distinctive groups, each one which guides to the next group, to a final outcome. They are: understanding a strategy; analyzing the position; planning a strategy; and implementing the strategy.

The first section called Understanding a Strategy, speaks of the first basic components that you should consider when looking to define a strategy and the basic components involved in thinking of solutions in a strategic way. Are you as an IT manager looking for short-term results or are you looking for future results? The IT manager should look at the employees and staff involved and seek to find which is better suited to assist in the process to find a solution which will achieve success. The IT manager can take a lot of waste and lost time out of the equation if you avoid guesswork and stick with the facts and statistical data. For success in the beginning stages and even in the final stages you must always keep reviewing the process to allow you to stay on track and not lose sight of the purpose of your project.

The second section called Analyzing the Position, speaks basically of knowing the audience you are getting an answer for, knowing what influences the results, knowing your audience needs, knowing your competitors, assessing yourself and your IT employee's skill and abilities and summarizing the data analysis of all of these factors combined. Every problem has at least one influencing factor this could be anything from monetary values to professional notoriety. What exactly is required of you by the company who has hired your IT firm or is the issue in-house? What are the audience expectations and limits? What are the aspects of competitors? What skills and knowledge does your IT solution team possess? When you combine all of this information together you will be better able to come up with selecting a plan of strategy that will assist your IT business in achieving success.

The third section called Planning a Strategy, speaks of the five stages to consider when completing a strategic planning project. Define the purpose, determine the advantages, set the boundaries of the project, choose the main areas to emphasize or which stand out, and lastly estimate the amount of money and time which will be allotted for the project solution. The IT manager should make sure that everyone involved in this project and solution is on the same page so that there is no conflict in either the project team camp or the customer's camp. Toward the end of this section there are two final steps in this process, to test to validate the strategies potential success that you have compiled and to make sure that communication has been successful and informed all of the people who need to be involved in this project.

The fourth and final section called Implementing Strategy is I would say the hardest step to complete. This step involves every single person at every level. When implementing a strategy there are several steps in the beginning of implementation which are setting priorities for the change: actually planning the change; assessing the potential risks involved in the change; and finally reviewing the targeted goals presented. Now we come to the hardest part that of motivating the staff. The majority of people I believe are scared of changes. Any internal or external movement in a corporate setting usually puts all employees on pins and needles. Usually if there is a change in the corporate situation that means that someone has a bright idea of improving something and that generally means that the improvement is going to benefit the company and not the employees.

You can bet that if the strategy benefits the shareholders that it matters not if you are a part of it or not so the only choice you have is to go along with it and hope for the best. After the motivational step there are three final steps to go through and they are monitoring the performance of the project after implementing, reviewing and analyzing the acquired data, and being flexible by looking at the data and if there are changes or alterations to be made to be successful then you must be open to suggestions and maybe even starting this process over again.

A B O U T   T H E   A U T H O R

I was born in an Appalachian county in central Kentucky.

Currently, I am involved in post-secondary education, bringing my ideas, concepts, philosophies, and experiences to my students. Students are the sole reason I do what I do. Currently, I am in the College of Business, Engineering and Technology where I am the Acting Chair and Assistant Professor of the School of Mathematics and Computer Science at Kentucky State University, where my specialties are Cybersecurity, Information Technology, and Network Engineering.

I am not a geographer nor a certified engineer. I see an issue and have looked at what I think to be a potential positive resolve.

NOTES

The End

www.ingramcontent.com/pod-product-compliance
Lightning Source LLC
Chambersburg PA
CBHW070643030426
42337CB00020B/4137